Project Report 41

Demonstration of the *Geodur* solidification/stabilisation system

F M Jardine
S T Johnson

CIRIA *sharing knowledge ■ building best practice*

6 Storey's Gate, Westminster, London SW1P 3AU
TELEPHONE 020 7222 8891   FAX 020 7222 1708
EMAIL enquiries@ciria.org.uk
WEBSITE www.ciria.org.uk

Summary

This report describes field trials and testing of the commercially available *Geodur* solidification/stabilisation system for processing contaminated soils and other industrial residues. It is one of a series of reports resulting from CIRIA's programme of case study demonstrations of remediation of contaminated land.

The treatment system, which uses proprietary additives to enhance performance, is based on cement stabilisation. The main trials involved two contaminated materials; additional trials were made on four different types of material imported from other sites. The mixed materials were transported to a nearby site and compacted as ground slabs. The report describes the field and laboratory work and presents the results of observations and tests on samples of the original and processed materials. The testing included determinations of compressive strengths of cube samples and specimens cored from the cast slabs at different ages and of chemical analyses and leaching. The report includes some of the findings of a subsequent project which examined the performance of the cast slabs over four years later. All had gained in strength below some surficial damage. Water entering subsurface and surface drains as leachate and as runoff appeared not to have been significantly affected by the stabilised material.

Demonstration of the *Geodur* solidification/stabilisation system

Jardine, F M and Johnson, S T

Construction Industry Research and Information Association

Project Report 41   © CIRIA 2000   ISBN 0 86017 841 2

Keywords
Contaminated land, solidification/stabilisation, technology demonstrations, metalliferous slags, waste recycling, contamination, industrial residues, stabilised strengths, chemical analyses, leaching tests

Reader interest	Classification	
Geotechnical, civil and environmental engineers, developers, regulators	AVAILABILITY	Unrestricted
	CONTENT	Record of field and laboratory work
	STATUS	Commissioned
	USER	Geotechnical, environmental and development professionals; regulators

Published by CIRIA, 6, Storey's Gate, Westminster, London, SW1P 3AU. All rights reserved. No part of this publication may be reproduced or transmitted in any form or by any other means, including photocopying and recording, without the written permission of the copyright holder, application for which should be addressed to the publisher. Such written permission must also be obtained before any part of this publication is stored in a retrieval system of any nature.

Acknowledgements

This report, a case study of the field trial of the *Geodur* solidification/stabilisation system at the Wath Recycling site near Rotherham, Yorkshire, forms part of Research Project 489 under the third phase of CIRIA's research programme about the remedial treatment of contaminated land. The objective of this programme is to promote safe, effective and economic remediation using the most appropriate technologies under UK conditions. The specific objective was to report the demonstration of the treatment process using six materials: two from the Wath site and four imported from other sites with a variety of metal and other contaminants. The report was written by Dr S T Johnson (formerly of CIRIA) and Mr F M Jardine of CIRIA. During the project, Mr M A Smith, who acted as environmental adviser to CIRIA, recommended the analytical testing programmme and provided commentaries on the analytical test results. Reference is made to – and quotations have been taken from – a TRL report to Highways Agency by Mr M J Board and Dr J M Reid.

This project report has been reviewed by the following persons:

Mr M J Board	University of Southampton
Mr P A Braithwaite	Ove Arup and Partners
Mr D L Edwards	ExSite
Dr J M Reid	Transport Research Laboratory
Mr M A Smith	M A Smith Environmental Consultancy.

CIRIA's research managers for the project were Dr S T Johnson and F M Jardine.

CIRIA is grateful to VHE Construction plc and Rotherham MBC for the opportunity to report the trial and for permission to use information gained from it in this report, and to Ove Arup and Partners and Aspinwall and Company (now Enviros Aspinwall) for arranging additional materials to be brought to the trial site. The case study would not have been possible without the help of Mr D L Edwards of VHE Construction plc. The authors gratefully acknowledge the work of Mr M J Board who carried out the further study with TRL and for the permission of Highways Agency and TRL to publish extracts from their report PR/CE/135/99.

CIRIA also gratefully acknowledges the following organisations for funding Phase III of CIRIA's "Programme on contaminated land":

- Department of the Environment, Transport and the Regions, Construction Directorate through the Partners in Innovation programme
- The BOC Foundation
- British Waterways
- The Environment Agency
- Highways Agency
- National House-Building Council
- Scottish Enterprise.

Further information about the *Geodur* system can be obtained from Dr W B Schwetlick, *Geodur* CIS AG, Oberallmendstrasse 20A, CH-6302 Zug, Switzerland and Mr D L Edwards, *Geodur* (UK) Ltd., Engine Lane, Shafton, Barnsley, S72 8SP, UK.

Executive summary

1. This report describes the field trial and demonstration of the *Geodur* system for solidification/stabilisation. In this project at Wath Manvers near Rotherham, Yorkshire, a special additive, cement, mix water and aggregate were added to two contaminated slags from the Wath recycling site and four wastes from elsewhere. Using readily available batching and mixing plant the raw materials were treated and transported to a nearby site at Wombwell. Here they were laid and compacted by roller into ground slabs. The slabs were left exposed to the weather, ie without surface protection.

2. The fieldwork was carried out in December 1994. The testing and analytical studies were in two phases. The first, the CIRIA study, was over a period of 17 months after the fieldwork; the second was during a study by the Transport Research Laboratory in 1998–1999.

3. The waste materials that were treated can be described as follows:

    A   Silty sand with ash and up to gravel-sized slag with some clay, with contamination by heavy metals and organics, classed as special waste.

    B   Silty sand with ash and up to gravel-sized slag with some clay, with contamination by heavy metals and organics, classed as hazardous waste.

    C   A lagoon-dried pulverised fuel ash from Eggborough power station, Yorkshire.

    D   A zinc-smelting slag.

    E   An intermediate (vertical retort) residual from zinc smelting.

    F   A contaminated silty clay with assorted masonry fragments and plastics from a former gasworks site on the Pride Park, Derby development.

4. The raw materials were brought to a small concrete batching plant together with about 18 per cent of limestone aggregate and mixed with 10 per cent of cement (proportions by dry weight of untreated material). The *Geodur* additive and water were mixed into the raw materials. Samples were taken at the time of mixing which were cast as cubes for subsequent strength, leaching and chemical testing. The mixed materials were transported by lorry to a prepared level subgrade at the Wombwell site; they were spread and compacted into slabs about 300 mm thick. Special drainage measures were incorporated into the site to collect leachate and runoff water for testing and to protect nearby surface watercourses.

5. A testing programme involving several laboratories was undertaken on samples of the untreated raw materials and at different elapsed times on cubes cast from the mixed materials and on cores drilled from the completed slabs.

6. All the wastes have concentrations of contaminants well above background levels. The treatment was effective in solidifying all of them, ie creating a bound material. The gain in compressive strength progressed at different but steady rates in the different materials, eg while there was little early strength for treated material D, its cube strengths (estimated from tests on core specimens) at ages between 91 and 1216 days (three to 40 months) were all over 9.7 N/mm^2. Even the weakest treated material F had strengths of at least 2.6 N/mm^2 after 1216 days.

7. The immobilisation of the contaminants is less clear. The procedure of the NRA leaching test is to comminute the treated material to finer than 5 mm size, ie destroying much of the physical occlusion of the waste and associated contaminants (although not affecting chemical bonding of contaminants within the cement matrix). Where reduced concentrations of particular contaminants were measured on leachate from treated as opposed to untreated material this could be partly attributed to the diluting effect of adding aggregate and cement (28 per cent of the dry mass of treated materials A and B). On the other hand greater concentrations of aluminium were measured in the test leachate from the treated materials, which is attributed to its being leached from the cement.

8. Nevertheless the analyses of leachate from the cast slab and from runoff water at the slab site showed it generally to meet UK drinking water standards other than for aluminium (in the first phase of study) and selenium and potassium (in the second phase TRL study) which were slightly over the limits.

9. It is not clear what the effect the *Geodur* additive had, as there were no control tests against which to compare (ie strength of materials treated with the same amounts of aggregate and cement but without the Tracelok™ additive). It is assumed, however, that it assisted stabilisation in the presence of potentially inhibiting organic compounds. Without knowing the chemical composition of the additive it is not possible to adduce its effect, if any, on fixation of the metals, nor to separate its contribution to resisting leaching or becoming part of the leachate itself.

10. At the small scale of the trial, costs have little meaning. Routine plant can be used and for larger volumes, there will be economies of scale with higher capacity equipment. Nevertheless, cement contents of the order of 10 per cent represent significant cost.

11. The demonstration of the *Geodur* system was successful in that contaminated materials were rendered to a condition in which they posed less threat to people and the environment. Some very poor quality wastes were transformed into viable, relatively strong and durable construction materials, that might be used, for example, as a subbase or lower pavement layer to a hardstanding or similar.

12. A general lesson for practice is that it is not sufficient only to test leachates for typical contaminants of concern, such as, for example, those of concern for human health by ingestion or inhalation, but that the analytical suite should cover all contaminants which might be of concern to surface waters and groundwater, eg aluminium.

Contents

Summary 2

Acknowledgements 3

Executive summary 5

List of figures 9

List of tables 9

Abbreviations 10

1   **Introduction** 11

2   **Objectives and approach** 13
    2.1   Background to the *Geodur* case study 13
    2.2   Objectives 14
    2.3   Approach adopted for the study 15

3   **The *Geodur* system** 17
    3.1   Introduction to solidification/stabilisation 17
    3.2   The *Geodur* solidification/stabilisation system 18

4   **Project programme** 21
    4.1   Project planning 21
    4.2   Project administration and organisation 22
    4.3   Principal organisations involved in the trial 23

5   **The field operations** 25
    5.1   The source site of material 25
    5.2   Preparation for the trials 26
    5.3   Plant and equipment 28
    5.4   Mobilisation 28
    5.5   Mix design 29
    5.6   Processing at the Wath site 30
    5.7   Slab construction 30
    5.8   Health and safety aspects 30
    5.9   Sampling and testing programme 33

6   **Main trial results: materials A and B** 35
    6.1   Engineering description of materials processed 35
    6.2   Strength testing 35
    6.3   Chemical analyses 36

7   **Subsidiary trials** 45
    7.1   Scope of the work 45
    7.2   Engineering and chemical descriptions 46
    7.3   Mix design considerations 47
    7.4   Compressive strengths 47
    7.5   Leaching tests (NRA method) on subsidiary trial materials 48

8	**Operational good practice**	**51**
8.1	Principles of good practice	51
8.2	Planning and organisation	51
8.3	Operations	52
8.4	Sampling and testing	53

9	**Discussion of results**	**55**
9.1	Contaminants in the untreated and treated materials	55
9.2	Leaching test results: leached inorganics	55
9.3	Leaching test results: leached organics	56
9.4	Permeability of the compacted treated material	56
9.5	Strength	56
9.6	Durability	56

10	**Concluding remarks**	**57**

References 59

LIST OF FIGURES

3.1	Requirements for solidification and stabilisation	17
3.2	Processes of stabilisation and immobilisation of contaminants	18
3.3	Chemical principles of the *Geodur* system	19
4.1	Initial programme for the trials	21
4.2	Project organisation	22
5.1	Plan of slabs cast at the Wombwell site	27
5.2	The batching and mixing operations	31
5.3	Loading the mixed material for transport to Wombwell	31
5.4	Slab construction	32
5.5	The completed slabs	32
8.1	Organisational structure	32

LIST OF TABLES

5.1	Pre-project chemical data	25
5.2	Averaged values from chemical analysis of materials A and B from the Wath site	26
5.3	Mix design for materials A and B from the Wath site	29
5.4	Feedstock and batching rates	30
6.1	Particle size distributions (average of three)	35
6.2	Cube sample strengths	35
6.3	Core sample strengths	36
6.4	Contents of five metals in samples of materials A and B (pre-project information)	36
6.5	Mean and comparison of means for chemical contents of untreated materials A and B	37
6.6	Results of NRA leaching tests on untreated materials A and B	38
6.7	Results of NRA leaching tests on untreated and treated material A	39
6.8	Total chemical contents of untreated and treated material A at the times of NRA leaching tests	40
6.9	Summary results of NRA leaching tests on untreated and treated material B	41
6.10	Summary results of total chemical content on untreated and treated material B at the time of NRA leaching tests	42
6.11	Results of chemical analyses on immersion water changed daily after three periods of immersion for cast cylinders	43
6.12	Analyses of runoff and percolation waters and downstream sediments at the site of the slab compacted from treated materials A and B	44
6.13	Inorganic contaminants in drain water samples collected in TRL study	44
7.1	Particle size distributions of subsidiary trial materials C, E and F	45
7.2	Chemical composition of material C	46
7.3	Chemical composition of material D	46
7.4	Treatment mixes of subsidiary trial materials	47
7.5	Compressive strengths of cube samples after two curing periods	48
7.6	Compressive strength of cores taken at different times after casting	48
7.7	Treated material C: results of leaching tests (NRA method)	48
7.8	Treated material D: results of leaching tests (NRA method)	49
7.9	Treated material E: results of leaching tests (NRA method)	49
7.10	Treated material F: results of leaching tests (NRA method)	50

Abbreviations

CDM	Construction (Design and Management) Regulations
CEN	European Standards Organisation
CIRIA	Construction Industry Research and Information Association
DNAPL	dense non-aqueous phase liquid
EA	Environment Agency
EQS	Environmental Quality Standards
HA	Highways Agency
l.o.i.	loss on ignition
MBC	Metropolitan Borough Council
NRA	National Rivers Authority
PAH	polyaromatic hydrocarbons
PFA	pulverised fuel ash
PSD	particle size distribution
t.e.m	toluene-extractable material
TRL	Transport Research Laboratory

1 Introduction

The Construction Industry Research and Information Association (CIRIA) case study demonstrations are part of a series of projects providing authoritative guidance about the remediation of contaminated land for the construction industry and its clients.

The demonstration projects and case studies form the third phase of CIRIA's geo-environmental programme on the problems of redeveloping contaminated land. The resulting reports extend the information and guidance already published by describing field-scale trials and full-scale remedial treatment projects. The series of case studies includes both *in-situ* and *ex-situ* remediation technologies and techniques. In this context, the term *remediation* covers all on-site activities from investigation and characterisation, through remedial treatment to monitoring and maintenance.

The research programme is jointly funded by industry and government on a project-specific basis. The principal government support is through the Construction Directorate, Department of the Environment, Transport and the Regions. The programme of case studies is included in the Department's Partners in Technology (now Partners in Innovation) programme for sponsorship of collaborative research on construction. Whenever possible, each project and case study report is guided by a CIRIA-appointed steering group of experts and representatives of the sponsoring organisations. When, as in this project, it was not possible to set up a steering group before the work started, CIRIA involved others as reviewers and in this case commissioned specialist advice.

The main objective of the CIRIA demonstration programme is to provide for the independent collection of data relevant to:

- the site-specific and wider applicability of the technologies included in the case studies
- the technical performance of these technologies and the implications of technology development for standard-setting
- regulatory acceptance and procedures
- the relative costs of the demonstrated technologies.

CIRIA's and the programme sponsors' intentions are to make the results of the studies widely known, in particular by publication of a project report on each study, in order to help in:

- promoting the wider uptake of selected technologies
- developing UK expertise in the application of a wide range of different remedial technologies
- increasing the level of scientific and technical understanding of land remediation technologies and techniques.

The project reports describe the treatment technology and present the results from the case study. As such they can be referred to when considering whether a technology is potentially suitable for use at a particular site. In accordance with CIRIA's practice, each report is subject to independent review. The reports, together with other site-specific information, can also be used to assist in the design and implementation of remedial strategies involving demonstrated technologies.

This report describes field trials and testing of a commercially available solidification/ stabilisation system for processing contaminated soils and other industrial residues. The treatment system, which uses proprietary additives to enhance performance, is based on cement stabilisation. The main trials involved two contaminated materials; additional trials were made on four different contaminated materials imported from other sites. The report describes the field and laboratory work and presents the results of observations and tests on samples of the original and processed materials. The tests were made over a period of 18 months on samples of the original mix and on cores taken *in situ* from the mass of treated material.

The trials took place in Yorkshire on the Wath Recycling site within the Wath Manvers reclamation area with the support of the contractor and technology vendor, VHE Construction plc and the local authority, Rotherham MBC. Others who were involved in the trials or in supplying additional materials for testing are listed in the report.

A subsequent study of the stabilised materials has been made by the University of Portsmouth under contract to the Transport Research Laboratory on behalf of the Highways Agency. The unpublished (at January 2000) contractor's report PR/CE/135/99 on that study is entitled *The effects of age on cement stabilised/solidified contaminated materials* (Board and Reid, 1999). Reference is made to the some of the results given in that report, insofar as they extend the original case study to four years and increase the analytical results, but because it was a separate research project with a different purpose, it is not described in detail here. Nevertheless the two reports should be mutually compatible and complementary.

2 Objectives and approach

2.1 BACKGROUND TO THE *GEODUR* CASE STUDY

One method for the management and control of unconsolidated contaminated materials is to solidify and stabilise them. This immobilisation method may be appropriate where the main contaminants in a soil are a number of different metals and metalliferous compounds. An opportunity to include an example of this type of treatment in the case study programme was offered to CIRIA by VHE Holdings. This permitted collaborative work, within the spirit of the framework protocol (Harris, 1996) for reporting demonstration projects, to obtain data and information from a full-scale site demonstration of a commercially available system for solidification/stabilisation. The specific treatment technique, the *Geodur* solidification/stabilisation system – at the time of the demonstration new to UK – is used in other European countries.

This opportunity arose because a major reclamation scheme was being undertaken by the parent company of the UK licence-holder for the *Geodur* system, VHE Construction plc. The demonstration was performed on the site of a former recycling and processing works, where the soil and other fill materials had been left in a contaminated state. The former site-owner had brought on to the site various waste materials to be reprocessed, including metals and other contaminated industrial waste. One of the activities, for example, was the recycling of nickel electrodes that had been used in the manufacture of oil products. Waste products had even been imported from the USA. It was known that concentrated copper solutions stored on the site had leaked into the ground over a long period.

The materials spread over the surface of this former recycling site and present in stockpiles are mixtures of soil and slag. The results of chemical analyses on samples previously taken from pre-existing stockpiles and the surface had been used to assess the material for disposal to a licensed landfill. On that basis, the stockpile material had been categorised as special waste, and the 150 mm thick surface layer as hazardous waste. This differentiation led to the terms used during the *Geodur* demonstration (and followed in this report) of materials A and B being from the stockpile and the surface layer respectively. In the event, further chemical analyses of materials A and B, but on samples taken from subsequently formed (ie new) stockpiles after excavation and preprocessing (screening), did not show sufficient difference in their constituents to distinguish them in terms of special or hazardous wastes. Nevertheless the identification of source (ie A or B) was maintained. More details of the engineering and chemical characteristics of the materials are given in Sections 5 and 6.

Following discussion and liaison with the local authority and the relevant regulatory bodies, including the waste regulator for the area and the (then) National Rivers Authority, the trial went ahead over a two-week period in December 1994.

Close to the former recycling site, the source of the contaminated material, is an area of land owned by VHE Holdings plc, in which are the offices of VHE and *Geodur* UK Limited. Part of this area was used for the placement of trial slabs of the treated materials.

This not only allowed ready access to the slabs for inspections and core-samples, but also provided security so that the treated material was undisturbed over the initial monitoring and test period of more than 18 months and until a later research project by TRL in 1998–1999.

In addition to *Geodur* UK Limited and CIRIA (on behalf of the funders of the case study demonstration programme), the following other organisations actively supported or took an interest in the trials: Rotherham MBC, the Regional waste regulatory authority, the former National Rivers Authority, Ove Arup and Partners and Aspinwall and Company (now Enviros Aspinwall) who provided subsidiary test materials.

The additional research project which was subsequently undertaken by the University of Portsmouth on behalf of TRL was funded by Highways Agency and VHE Construction Ltd. As well as instituting a new programme of sampling and mechanical and chemical testing on the treated materials of the cast slabs, the TRL work used the data obtained in the original trial that was supplied by VHE and CIRIA.

2.2 OBJECTIVES

CIRIA's objectives in reporting this case study demonstration of the *Geodur* immobilisation system are as follows.

1. To make independent information available on this particular system for solidification/stabilisation

2. To show, where appropriate, the generic applicability of this type of treatment for the remediation of contaminated land or management of contaminated materials

3. To review the potential of the *Geodur* solidification/stabilisation system for processing contaminated soils and industrial by-products in respect of:
   - environmental impact and protection criteria
   - the technical performance of the system to immobilise predominantly inorganic contaminants against leaching, using the NRA leaching test as the standard
   - the range of application of the system in terms of contaminants treated and the characteristics of material to be treated
   - operational and management requirements
   - third-party concerns about the processed materials
   - relative costs and benefits of the system compared with other options.

VHE Construction's objectives in carrying out the demonstration were:
   - to gain experience with the *Geodur* system
   - to test the system in full view of the industry and its regulators
   - to demonstrate the ability of the *Geodur* system to process contaminated soil and other materials
   - to assess the potential for the processed material to be re-cycled and used as a engineering material, eg in construction of hardstandings.

The objectives of the TRL study were:

- to assess the variation in physical and geochemical properties of cement stabilised/solidified materials with time
- to understand the leaching behaviour of the materials and identify mechanisms for the leaching of contaminants
- to identify potential applications for the reuse of stabilised/solidified materials for highway construction.

2.3 APPROACH ADOPTED FOR THE STUDY

The case study was designed as a demonstration of an on-site, *ex-situ* process of treatment. The process is what is termed solidification/stabilisation. The demonstration project was self-contained in the sense that it was independent of any overall remediation strategy for the former recycling site, which was the source of the materials. Thus it was not, nor was it intended to be, a case study of a remediation project.

In support of the fieldwork there was an appropriate but, because of time and financial constraints, limited sampling and testing programme. This included strength tests, chemical analyses, and leaching tests. The chemical analyses and leaching tests were carried out at the laboratories of Clayton Environmental Consultants, and the strength tests were carried out by SGS Laboratories. An expert environmental chemist, Mr M A Smith, advised the participants in the demonstration on chemical testing and on the interpretation of the results (see Section 3).

The contaminated materials were excavated from the surface and stockpiles of the former recycling site (the Wath site), and processed on that site. They were then transported to the nearby site at Wombwell, placed and spread on a prepared ground surface and compacted by roller. In effect, cast slabs of stabilised soil, they were left exposed to the elements for five years. Over the 18 month period covered by this report, samples were taken from the slabs by core drilling for leachability testing, compressive strength determination, and chemical analysis. The general condition of the surface of the treated material was regularly inspected and photographed. In 1998, for the TRL research project, the University of Portsmouth took further samples including drilled cores and water samples for testing and analysis.

In addition to contaminated soil materials A and B, it was possible to include within the demonstration small quantities of one other contaminated soil and three industrial waste by-products and residues, ie:

- material C – a PFA
- material D – a zinc-smelting slag
- material E – an intermediate (vertical retort) residual from zinc smelting
- material F – a contaminated soil from the made ground of a gasworks site.

These materials are described in Section 7 of this report.

3 The *Geodur* system

3.1 INTRODUCTION TO SOLIDIFICATION/STABILISATION

The following definitions of solidification and stabilisation are taken from CIRIA Special Publication 107, *Remedial treatment for contaminated land, Volume VII: Ex-situ remedial methods for soils, sludge and sediments* (Harris, Smith and Herbert, 1995 b).

Stabilisation involves adding chemicals to the contaminated material to produce more chemically stable compounds or constituents, for example by precipitating soluble metal ions out of solution. It may not result in an improvement in the physical characteristics of the material, eg it may still remain a relatively mobile sludge, but the toxicity or mobility of the hazardous constituents will have been reduced by the process.

Solidification involves adding reagents to contaminated material to reduce its fluidity/friability and prevent access by external mobilising agents, such as wind or water, to contaminants contained in the solid product. It does not necessarily require a chemical reaction to take place between contaminants and the solidification agent, although such reactions may take place depending on the nature of the reagent.

The requirements for successful stabilisation and solidification are shown in Figure 3.1.

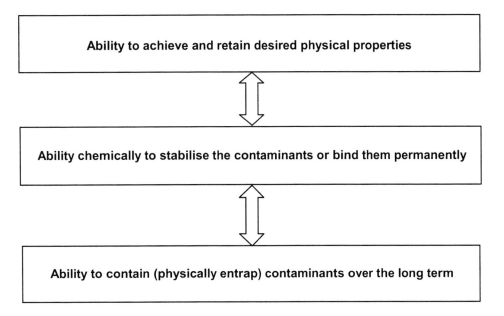

Figure 3.1 *Requirements for solidification and stabilisation*

In a process involving cementitious or pozzolanic reactions, contaminants may be physically or chemically bound or encapsulated within a stabilised, often solidified, mass. Chemical agents can be added to promote reactions that will enhance the stabilisation and achieve more effective immobilisation (to leaching) of the contaminants. The aim is to reduce both the short- and long-term leaching potential of the processed material. Stabilised contaminated materials are typically disposed to landfill or to some other suitable containment.

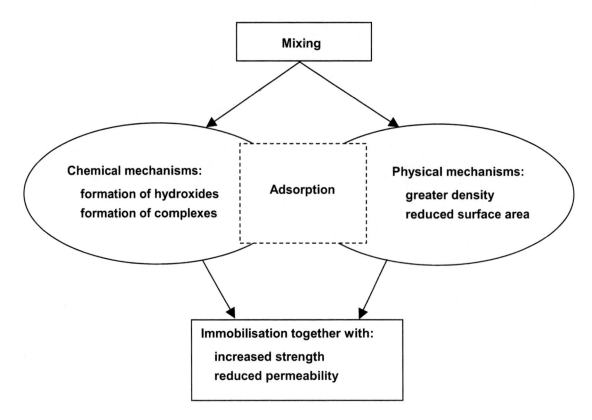

Figure 3.2 *Processes of stabilisation and immobilisation of contaminants*

Increasingly, however, these materials are being considered for possible reuse. Figure 3.2 is a generic flow diagram for the various activities and processes which contribute to stabilisation and immobilisation.

For most solidification and stabilisation systems the target contaminant group is inorganic, including radionuclides. Techniques using cementitious binders have not proved effective where there is a substantial organic content, particularly of semi-volatile products and pesticides (Harris, Smith and Herbert, 1995 b). Above trace levels, general experience is that organic or humic matter significantly reduces the ability of cement-mixed materials to solidify. Attributed to the *Geodur* system (see Section 3.2) is a tolerance of organic contaminants and an ability to reduce their availability. This makes *Geodur* (and its like) a viable remediation option where there is some organic contamination mixed with predominantly metallic or other inorganic compounds.

A fuller explanation of solidification and stabilisation and its variations is given in CIRIA Special Publication 107 (ibid).

3.2 THE *GEODUR* SOLIDIFICATION/STABILISATION SYSTEM

The *Geodur* system for solidification/stabilisation of contaminated material comprises mechanical processes of mixing (and often of compaction) with chemical and physical changes to the material constituents. These changes are effected or enhanced by a proprietary additive, usually in combination with cement and mixed into the contaminated soil or industrial waste material. The additive, in association with the cement, promotes additional chemical bonding to control the availability of the contaminants.

Only a general description of the system is given below as the exact composition of the *Geodur* additives are subject to commercial confidence. What is known, however, is that the additive is synthesised from a number of compounds, some of which are commercially available, others are protected and produced under licence for the Swiss company *Geodur* CIS, the originator of the system.

Mixed with a cementitious binder, the *Geodur* system is designed to promote:

- the formation of complexes between contaminants and *Geodur* ligands
- chemical molecular binding of inorganic and organic contaminants with *Geodur* components.

These effects are additional to the formation of hydroxides and crystallisation, the increase in density, and the reduction of surface area, when cement is used to bond with the contaminants and solidify the mixture. Figures 3.2 and 3.3, adapted from literature provided by *Geodur* CIS, illustrate some of these effects.

Geodur's general approach to treating contaminated materials is described as a series of steps that start with problem definition and setting the objectives in terms of the requirements of the solidified/stabilised material.

Geodur effects

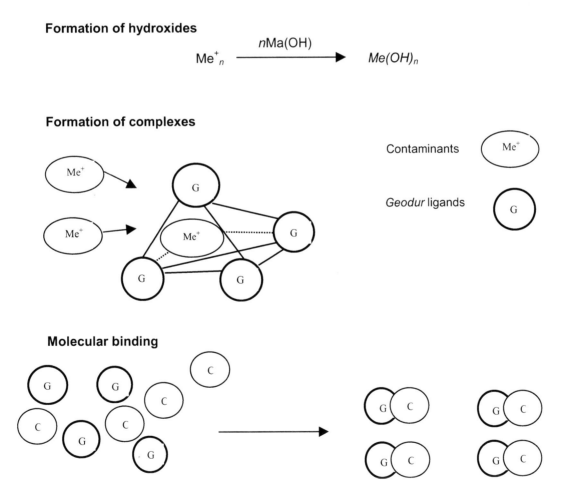

Figure 3.3 *Chemical principles of the* Geodur *system*

Treatability tests, carried out in the laboratory on samples of the contaminated soil or waste, are essential in order to:

- assess the applicability of the *Geodur* system to treat those materials
- optimise the mix design and the form the additive is to take.

The composition of the additive can be varied to suit the particular application and contaminants.

A number of projects have been carried in other European countries by *Geodur* over the last ten years using this system. One of the most common uses is in transforming difficult residues and waste materials so that they are easier to handle. By reducing and controlling any release of contamination from the processed material, in some countries the wastes may be reclassified as being of a lower hazard, such that disposal costs can be dramatically reduced. Some of the usages include the potential recycling of the treated material.

4 Project programme

4.1 PROJECT PLANNING

When the opportunity arose for this set of trials to be the first project in the CIRIA case study programme, there was not enough time to set up CIRIA's usual systems, such as a steering group to approve the inclusion of the study in the programme and advise the project team beforehand on the content of the work. Rather, in order to seize the chance of reporting this technology trial, arrangements were made to define and agree roles and responsibilities and call upon an expert for advice. The arrangements were able to be made quickly for the following reasons:

- the main contractor for the adjacent works was also the vendor (ie the UK licensee for the *Geodur* system) who therefore had in place the necessary site infrastructure, eg cabins, site engineering services, storage, access to a materials testing laboratory, transport and other plant, as well as ready access to the site
- the local authority – Rotherham MBC – who were already involved as the client for the main works were supportive of the trials
- the contractor/vendor had a site nearby, ie Wombwell, where the treated materials could be placed as a series of compacted ground slabs. Thus access for inspection and subsequent testing was guaranteed for at least one year.

The initial programme for the planning, implementation, testing and reporting stages of this case study trial is given in Figure 4.1.

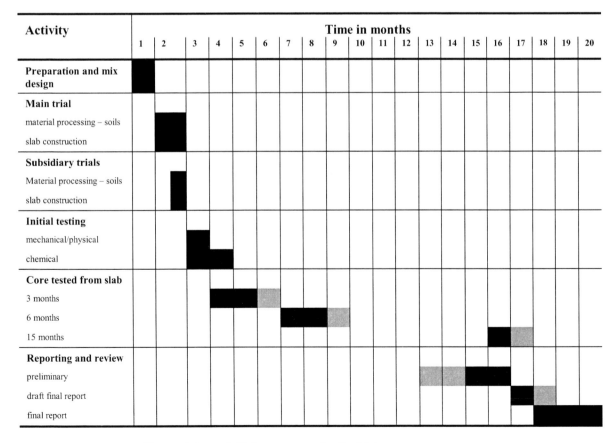

Figure 4.1 *Initial programme for the trials*

4.2 PROJECT ADMINISTRATION AND ORGANISATION

The work of the trials was managed by *Geodur* (UK) Ltd, who also undertook the relevant administration and liaison with the authorities and regulators. Those organisations and persons who were informed of the trials or in some other way involved in them were as follows:

- *Geodur* UK (VHE Construction plc), the UK licensee for the *Geodur* system
- CIRIA staff and sponsors of the case study programme, involved in preparing the case study report, peer review and dissemination of the findings
- Mr M A Smith, CIRIA's environmental consultant for the project
- Rotherham MBC, local authority
- South Yorkshire Waste Regulatory Authority, relevant waste regulator
- National Rivers Authority, relevant water quality regulator
- *Geodur* CIS, vendor of the solidification/stabilisation system
- Clayton Environmental Consultants, analytical chemical testing laboratory
- SGS UK Ltd., materials testing laboratory
- Ove Arup and Partners, subsidiary test material supplier
- Aspinwall and Company, subsidiary test material supplier.

The principal organisations involved and their respective roles and responsibilities are discussed in more detail in Section 4.3.

A diagram illustrating the roles and responsibilities and interactions between the various parties is presented in Figure 4.2.

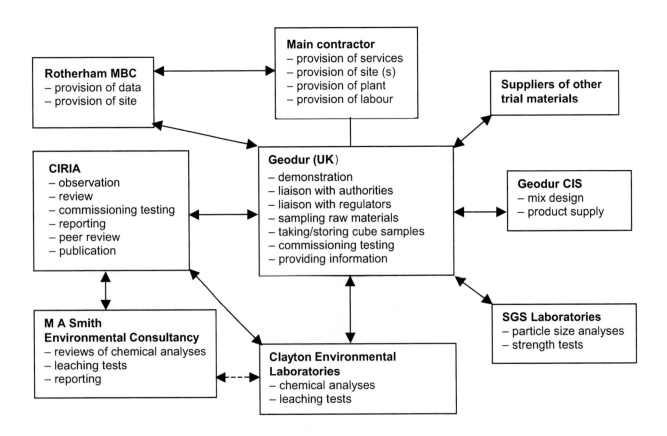

Figure 4.2 *Project organisation*

4.3 PRINCIPAL ORGANISATIONS INVOLVED IN THE TRIAL

This section lists the principal organisations and individuals involved in the *Geodur* demonstration and outlines their roles and responsibilities.

Geodur (UK) Ltd.

A wholly owned subsidiary of VHE Holdings plc. VHE Construction were the main contractors for Rotherham MBC on the site adjacent to the Wath recycling site. VHE Construction plc provided the area of land for construction of the slabs of treated materials; and they arranged for the necessary plant and site preparatory works. *Geodur* (UK) Ltd., as the technology contractor, carried out all the planning, hired the specialist plant and made arrangements for testing. *Geodur* (UK) Ltd. also liaised with *Geodur* CIS and with the authorities and regulators.

Construction Industry Research and Information Association (CIRIA)

CIRIA provided the means for these trials to be reported through the case study demonstration programme, for some of the testing to be undertaken, for the results to be independently reviewed, and the report published. CIRIA staff members were present during the main trials and were instrumental in commissioning some additional testing and employing an environmental consultant to interpret the chemical and leaching test results.

Geodur CIS AG

The Swiss company which is the holder of the patents on the various additives under the *Geodur* process. They provided product and technical support to the trials, as follows:

- laboratory treatability studies
- mix design and optimisation
- on-site monitoring and advice during the trials
- post-process sampling and analysis in their own laboratories for product quality assurance and performance.

Rotherham MBC

The local planning authority, the main client for all the reclamation work in the area in relation to the former colliery and railway marshalling yards at Wath Manvers, and the local environmental health regulator and inspector. Staff of Rotherham MBC's engineering, planning and environmental health departments, at various times, visited the trials and inspected the constructed slabs.

Material suppliers

A number of organisations, in particular Ove Arup (Midlands) and Aspinwall and Company, arranged for other materials to be brought to the site. These materials included various industrial by-products and residues and a contaminated soil (made ground) for the subsidiary trials.

Clayton Environmental Consultants

Clayton Environmental Consultants were sub-contracted jointly by CIRIA and *Geodur* (UK) Ltd. to carry out the chemical analytical testing on the unprocessed and processed materials and the tests for water quality both on the surface water run-off from the main slab and for any infiltration through the slab.

M A Smith, independent environmental consultant

The consultant assessed and inter-preted the analytical chemical test results, ie the testing done by Clayton Environmental Consultants on the various samples taken at the time of the trial demonstration and on the cores taken at intervals from the slabs.

Transport Research Laboratory (TRL)

TRL commissioned by the Highways Agency in September 1995 to establish which existing methods of processing contaminated land are economically and environmentally acceptable for highway works. As part of that commission, TRL instituted further research in association with the University of Portsmouth into the performance and properties of the cast slabs at Wombwell. This work was carried out by Mr M J Board initially with a CASE studentship there with support from VHE Construction plc and subsequently under a TRL contract.

5 The field operations

This section of the report describes the trials carried out on materials from the former Wath recycling operation and other imported materials. The trials involved work at the Wath site itself and at another place, Wombwell, where the treated materials were cast into slabs.

5.1 THE SOURCE SITE OF MATERIAL

The recycling operations on the Wath site had left contaminated soil and slag and soil/slag mixtures spread over the surface of the site and in stockpiles. In the terminology for the engineering description of soil, the soil/slag mixture would be described as a silty sand with ash and sand- and gravel-sized slag and some clay.

The contamination had been confirmed through investigations commissioned by Rotherham MBC. Analytical results provided by the local authority from earlier sampling of the site focused on the five metals, the range of measured concentrations of which are shown in Table 5.1, and which by their nature and very high concentrations could be a significant risk to human health and the environment.

Table 5.1 *Pre-project chemical data*

	Nickel	Copper	Lead	Cadmium	Zinc
Concentration range (mg/kg)	15–9000	35–494 000	25–41 000	<5–300	25–86 000
Number of samples	38	38	8	8	38

The residual soil contamination made reclamation of this site different from that being carried out by the local authority on the surrounding land. The mixtures of soil and slag in existing stockpiles and forming the surface layer of the site had been classified as either special or hazardous waste (material A and B respectively) for the purpose of disposal to landfill, considered the most practical remedial treatment strategy.

A solution appropriate in some circumstances for this type of problem is controlled disposal of contaminated material to an on-site containment area or cell. This gives the possibility of applying an *ex-situ* treatment to the material. Where viable, *ex-situ* treatment with the purpose of immobilisation of the contaminants, can be used to:

- render a contaminated soil or industrial residue to a condition sufficient to reduce the risks it presents during transport and after disposal at a licensed landfill to an acceptably low level
- produce an engineering material for reuse, eg as a sub-base or structural fill, with adequate short- and long-term resistance to leaching, in addition to the required engineering properties of strength, durability, etc.

How practical an approach immobilisation is depends on the nature and amounts of the contaminants present. For the Wath materials, the main contaminants were metals. There was, however, a substantial content of organics and some inorganic compounds that would tend to be adverse to cementitious stabilisation of the material. Table 5.2 lists for the two materials A and B average values of the concentrations of their different constituents and other test results. The results are from samples of the two material types taken at the start of the project and so are not the same as those in Table 5.1 by which the waste classification was derived. The values in the table are given in this way here as a broad indication of the typical chemical characteristics of the untreated soil/slag and the context in which to set the performance of the solidification/stabilisation process by the *Geodur* system. For more information on the chemical analyses made in relation to the trial materials see Section 6. The relatively large amount of organics is important to note as organic materials can often delay or even prevent cementitious mixtures from solidifying and gaining strength.

Table 5.2 *Averaged values from chemical analysis of materials A and B from the Wath site*

Determinand	Value (mg/kg or ppm)	Determinand	Value (mg/kg or ppm)
Loss on ignition @ 105°C (%)	1.6	pH	7.9
Aluminium (Al)	14155	Lead (Pb)	1232
Arsenic (As)	57	Mercury (Hg)	1.2
Cadmium (Cd)	332	Nickel (Ni)	4000
Chloride (Cl)	71	Phenols	1.7
Chromium (Cr)	255	Sulphate	5217
Copper (Cu)	3791	Sulphide	20
Cyanide (Cn)	3.6	Zinc (Zn)	1292
Total PAHs	80	t.e.m.*	13745

* t.e.m. = toluene-extractable material

The chemical and physical character of these materials and the environmental performance of the processed materials, including the ability of the *Geodur* system to cope with organic fractions, are considered in more detail in the results of the demonstration presented in Section 7.

5.2 PREPARATION FOR THE TRIALS

5.2.1 At the Wath site

The stockpiled material was excavated from its initial locations, put through a 63 mm screen to remove oversize lumps as a pre-processing, and re-stockpiled adjacent to mixing plant brought in for the demonstration. The material forming the ground surface of the site was then scraped to a depth of about 150 mm. A proportion of this excavated material was also put through the screens and placed in a separate stockpile. The two stockpiles for the demonstration were thus formed of materials A and B respectively. The amount of material in each stockpile was of the order of 200 tonnes.

Hard-standings remaining from the former recycling works were used to provide a stable foundation for the plant and equipment of the trials. They also helped in the control of materials dropped from conveyors and of surface water. The stockpiles were so positioned that loader movements off the hard-standing areas, which might have raised dust, were kept to a minimum.

The access into the site was suitable for lorries and ordinary road traffic. This was regularly graded and maintained during the demonstration period to give easy access for all necessary vehicles.

Water was provided from the mains supply at the site. A mobile generator provided the electricity for the site huts and electric power tools. The batching plant had its own self-contained power supply.

5.2.2 Preparation of slab area at Wombwell

One aim of the trial of the *Geodur* solidification/stabilisation system with contaminated materials was to examine if the products could be compacted and laid for reuse as an engineering material. The arrangements were made therefore to cast a series of ground slabs that would be available for inspection, sampling and testing over a period of at least a year. The site chosen for the ground slabs was at Wombwell on land owned by VHE Construction plc. It is within a few miles (a lorry travel time of about 15 minutes) of the Wath site where the solidification/stabilisation treatment took place. As well as the advantage of easy access, this location meant that the slabs would not be disturbed over the duration of the trial.

The location for slab construction was graded level in order to provide space for a slab about 25 m × 35 m in plan from the treated contaminated soil of the Wath recycling site and for a number of smaller (8–12 m long × 4 m wide) subsidiary slabs. A Visqueen™ (polyethylene) membrane was laid under the main slab. Drainage channels were constructed separately to catch any leachate which might come through the slab and the run-off from the surface of the slab. Figure 5.1 is a plan of the casting site and shows the position of the main slab of processed materials A and B, the drainage system for this slab and the positions of the smaller slabs cast from the other four processed materials C to F.

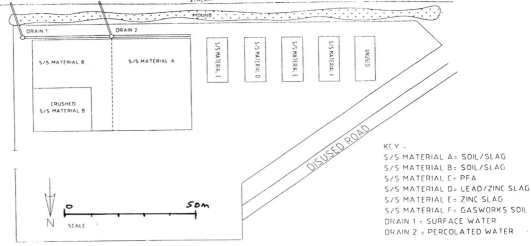

Figure 5.1 *Plan of slabs cast at the Wombwell site*

5.3 PLANT AND EQUIPMENT

5.3.1 Batching plant

A Belmix 50 concrete batching plant was used to mix the trial materials for the demonstration. This is a mobile, screw-auger continuous-feed mixer, with a maximum capacity of 100 t/hr. After the batching plant had been brought to the site and set up, two days were allowed for calibration before the planned start of the demonstration. A further day was needed to modify the single hopper of the batching plant so that it could handle not only concrete aggregates, but also the contaminated materials and industrial residues to be processed, and to calibrate the plant accordingly.

5.3.2 Other plant and equipment

The other items of plant and equipment used at the batching site included:

- a 1 t capacity front loader
- a power-screen for pre-processing the contaminated soil/slag materials A and B, prior to stockpiling ready for processing
- temporary accommodation and storage
- compressor for power tools
- cube moulds for concrete test cubes.

The plant used at the Wombwell casting site included:

- a Caterpillar 955 'four-in-one' tracked shovel
- a towed 1800 mm wide vibrating roller Vibroll 72T.

5.4 MOBILISATION

Preparation and mobilisation for the demonstration included:

- briefing the chosen testing laboratories on the scope and nature of the tests required (chemical analyses, leaching tests and cube and cylindrical crushing tests or unconfined compression tests on specimens from core samples) and about the materials to be tested
- providing a rudimentary site laboratory for trial mixes
- obtaining cube and cylinder sample moulds and Kango™ hammer equipment to compact the cubes (a vibrating table had been originally specified and would be preferable, especially for a site involving greater quantities)
- obtaining and stockpiling 10–20 mm and 20–40 mm aggregates
- ordering supplies of cement
- provision of safety equipment, masks, gloves and overalls appropriate for the work and materials being handled
- preparing safety guidance for all site workers and visitors
- liaising with local authorities and regulators.

5.5 MIX DESIGN

It would be normal practice to develop a mix design through feasibility studies and treatability testing, but it was not possible here because:

- the time available to plan the trial was short
- samples of the case study materials were not readily available to carry out a formal programme of treatability testing.

A small amount of testing was carried out on site by the *Geodur* CIS representative to determine appropriate mixes for the trial materials. Small cylindrical samples of the processed material were also taken by *Geodur* CIS for QA/QC and performance testing.

Based on this test and design work, a mix was decided which, in the experience of the *Geodur* CIS representative, would meet the objectives for the case study. It had the following components:

- the *Geodur* additive
- cementitious binder
- water
- aggregates
- the material to be processed.

***Geodur* additive**: the additive Tracelok™ was recommended for the trial materials by *Geodur* CIS. This is a general-purpose product within the *Geodur* system and is claimed to be tolerant of, and able to treat, a wide range of inorganic and organic contaminants at various concentrations and mixes.

Binder: the cementitious binder used in the demonstration for all the processed materials was an ordinary Portland cement supplied by Castle Cement.

Aggregates: 10–20 mm and 20–40 mm graded crushed sandstone aggregates were added in equal volumes to the contaminated soil/slag to improve the engineering characteristics of the compacted material and to give the raw mix a more manageable consistency and strength.

The mix design for materials A and B is given in Table 5.3. The moisture content of the excavated, pre-processed and stockpiled materials after combination with aggregate was approximately 17 per cent. The final moisture content of the mix was to be 18 per cent so only a little water was added in the mixing.

Table 5.3 *Mix design for materials A and B from the Wath site*

Mix design	Proportions by mass
Water:cement ratio (overall)	1.79:1
Water content of soil/slag material and aggregate	17 %
Geodur additive	0.15 %
Water (added)	0.85 %
Cement	10 %
Aggregate	18 %
Contaminated soil/slag	54 %

5.6 PROCESSING AT THE WATH SITE

The materials were continuously batched and mixed on the Wath site using the Belmix 50 batching plant at the production rate set for the trials of 56.8 t/h. In all, about 400 t of material was processed for the main demonstration slab, of which 180 t was material A and 100 t (wet weights) was material B. The mix proportions were converted to rates of feedstock into the batching and mixing plant as shown in Table 5.4.

Table 5.4 *Feedstock and batching rates*

Rate of flow of feedstock (material A or B plus aggregate)	14.22 kg/s	
Water content of mixture	17% = 0.17	
Rate of flow of dry feedstock		=14.22 x (1/(1.17) =12.16 kg/s
Rate of *Geodur* additive at 0.15% of dry feedstock	= (0.15/100) x12.16 x 60 = 1.09 l/min	
Moisture in flow of feedstock	= 0.17 x 12.16 = 2.067 kg/s	
Cement content	10%	
Cement addition rate		1.216 kg/s
Required final moisture content	18%	
Required total water		2.407 kg/s
Water to be added in mixer, less *Geodur* additive	= (2.407 – 2.067 – 1.09/60) = 0.34 kg/s	
Total flow of mixed material		15.783 kg/s

The batching and mixing plant can be seen in Figure 5.2 and Figure 5.3 shows the loading conveyor.

5.7 SLAB CONSTRUCTION

The processed material was loaded direct from the batching plant into lorries and transported to the prepared area at Wombwell. Typically the transport time was 15 minutes. The materials were spread and levelled using the tracked shovel and compacted by the vibrating roller pulled by the tracked shovel. The main slab was made of 280 t of processed materials A and B. It was 25 m × 35 m in area and 300 mm thick. The four subsidiary slabs, each made from approximately 12 t of processed materials C to F, were also 300 mm thick but only some 8–10 m long × 4 m wide.

Figure 5.4 is a photograph taken during construction of the slabs and Figure 5.5 shows the slabs when completed.

5.8 HEALTH AND SAFETY ASPECTS

The main health and safety concern was for those persons directly handling the contaminated materials and operating the batching plant. Personal protective clothing (appropriate dust masks, safety glasses or goggles and disposable overalls) were required to be worn by all personnel who would be in direct contact with contaminated material. These were additional to the compulsory wearing of hard hats and protective boots. During the time of the trials, the weather was mainly dry with only light winds. Generally, therefore, dust was not a problem, except in the immediate vicinity of the batching plant when it was being loaded with the drier materials.

Figure 5.2 *Batching and mixing operations*

Figure 5.3 *Loading the mixed material for transport to Wombwell*

CIRIA PR41

Figure 5.4 *Slab construction*

Figure 5.5 *The completed slabs*

CIRIA PR41

5.9 SAMPLING AND TESTING PROGRAMME

The following schedule for sampling and testing was established for both untreated and processed materials.

5.9.1 Sampling

Wath recycling site: untreated materials A and B

One sample of the feedstock was taken each hour and, in addition, further samples were taken before and after each sampling of the processed mix to make test cubes. Samples were labelled by time and date and stored in closed-lid containers. Samples were taken on each day of casting.

Wath recycling site: processed material A and B

Test cubes were made up in three sets of eight per day (ie a total of 24 cubes per day). Two cubes from each set of eight were cured under the same (ambient) conditions as the slabs for testing at 28 days, the rest were cured as is standard for concrete (ie underwater at constant temperature). All tests were carried out by a NAMAS-accredited laboratory.

Standard tests for crushing strength were made on one sample from each set of eight at ages of one, seven and 28 days, and one year. The remaining two cubes from each set were retained in long-term storage for inspection, sectioning or testing in relation to such aspects as weathering, long-term durability and permeability.

Subsidiary trials: untreated materials C to F

Samples were taken at approximately one-hour intervals during casting and before and after each set of cube samples of the processed mixes.

Subsidiary trials: processed materials C to F

Twelve cubes were made from each of the test materials supplied for trial. From each batch of 12 cubes per trial material: two cubes were cured at ambient conditions for testing at 28 days; the remaining cubes were cured as standard for concrete, with two cubes to be tested at each age of one, seven and 28 days, and one year. The remaining two cubes were retained for other tests or inspection. Note that not all the tests were carried out.

TRL research project

Some 40 100 mm dia cores were taken from the slabs in March 1998. Water samples were taken from the sump of the surface run-off system. The leachate at the base of the main slab was also sampled.

5.9.2 Testing

Characterisation of materials to be processed

The following tests were carried out on all the materials to be processed:

- moisture content
- bulk density
- particle size distribution (PSD) of untreated material
- leachability (NRA test)
- leachability (Draft CEN test)
- bulk chemical analysis for organics, fats, metals, heavy metals, volatile and semi-volatiles.

Cement and aggregate added in the processing

Tests were carried on the cement and aggregate used in the processing to determine their chemical characterisation and leachability.

Subgrade soil at Wombwell

One sample of the subgrade soil at the Wombwell site, ie the formation material on which the slab was laid, was taken for description and bulk chemical analysis.

Processed material

Samples of the processed materials were taken after mixing and cast into cubes for testing as follows:

- crushing strength (2)
- bulk density measurement
- leachability (NRA)
- leachability (Draft CEN test)
- permeability/porosity
- long-term durability/chemical compatibility
- chemical character of curing water; before and at 28 days.

Samples from the cast slabs

Core samples taken by diamond drilling were taken from the cast slabs for testing as follows:

- coring at three, six and 15 months
- leachability (NRA)
- leachability (Draft CEN test)
- bulk density measurement
- compressive strength
- leached water quality
- run-off to watercourse, water quality
- other environmental monitoring for compliance, as appropriate and required.

Subsequently other core samples were taken in the TRL study.

6 Main trial results: materials A and B

6.1 ENGINEERING DESCRIPTION OF MATERIALS PROCESSED

Materials A and B are mixtures of soil and slag and can be described as sandy ash and slag gravel. Particle size distributions for materials A and B are given in Table 6.1. The natural moisture content of both materials A and B was measured at 17 per cent.

Table 6.1 Particle size distributions (average of three)

Particle size	Percentage passing (by dry weight)	
	Material A	Material B
5.6 mm	66	77
2 mm	35	44
500 μm	12	19
212 μm	5	7
< 63 μm	0	1

6.2 STRENGTH TESTING

Compressive strength determinations, in accordance with BS 1881: Part 120: 1993, on cubes made from the processed trial materials and on cores taken subsequently from the cast-in-place and roller compacted slabs were as summarised below.

6.2.1 Cube samples

The cube strengths at 28 days and 12 months are given in Table 6.2.

Table 6.2 Cube sample strengths

Processed material	At age of 28 days (or 1 month)		At age of 12 months	
	Cube strength (N/mm^2)	Density (kg/m^3)	Cube strength (N/mm^2)	Density (kg/m^3)
A	3.25	2100	7.75	2040
B	7.25	1940	12.0	2030

6.2.2 Core samples

Cores were taken from the cast slab at Wombwell at three, six, 12 and 15 months. For the Wath recycling materials, the results of unconfined compression tests on specime trimmed from the cores are presented in Table 6.3. In this table, these strengths have been converted to equivalent cube strength.

Table 6.3 *Core sample strengths*

Processed material cored from slab	Age (mths)	Core strength (N/mm^2)	Equivalent cube strength (N/mm^2)	Density (kg/m^3)
A	3	12.5	12.0	2100
	6	7.7	7.0	2040
	15	10.5	10.0	1990
(4 tests)	40	6.9–11.0	6.6–10.6	2077–2126
B	3	6.5	6.0	1940
	6	5.5	5.0	2030
	15	8.0	7.5	2010
(3 tests)	40	9.1–17.6	8.6–16.8	2091–2134

6.3 CHEMICAL ANALYSES

This section presents the results of chemical analyses on untreated materials A and B. Most of the tables in this section are summarised data: for reasons of space individual analytical results have not been included in the report.

6.3.1 Pre-project information

Rotherham MBC provided VHE Construction and *Geodur* (UK) with information on the chemical content of materials A and B from samples taken in early 1994 on the Wath recycling site. These data were used by Rotherham MBC to categorise the materials for their probable disposal to licensed landfill as the most likely remediation strategy for the site. Chemical analyses concentrated on five key contaminants: cadmium, copper, lead, nickel, and zinc. Amalgamated results of determinations of the contents of these metals are given in Table 6.4 (the same results as in Table 5.1).

Table 6.4 *Contents of five metals in samples of materials A and B (pre-project information)*

Metal	Content in mg/kg			Number of samples
	Maximum	Minimum	Mean	
Cadmium	300	< 5	61	8
Copper	494 000	35	42 225	38
Lead	41 000	25	10 341	8
Nickel	90 000	15	8943	38
Zinc	86 000	25	8202	38

Chemical analyses of the untreated materials

A full range of chemical determinands were specified to be analysed for both materials A and B. Mean values and ranges of results are given in Table 6.5. On the basis of these results, the contents of the analysed constituents of materials A and B as sampled and processed in the demonstration are generally not significantly different at the 95 per cent confidence level.

Inspection of the contents of the five key metals in Tables 6.4 and 6.5 shows a marked difference between the pre-project and demonstration results respectively. With the exception of cadmium, all the measurements of metal contents from the trial materials are substantially less than those of the pre-project analyses. The reasons for the differences are that not only were the sampling locations different, but also that trial materials A and B had been screened and stockpiled prior to being sampled. Thus Table 6.5 represents the more uniform chemical contents of the untreated materials immediately prior to being put through the *Geodur* process.

Table 6.5 *Mean and comparison of means for chemical contents of untreated materials A and B*

Determinand	Unit	Untreated material A			Untreated material B			Significance level (%)
		Mean	Min.	Max.	Mean	Min.	Max.	
Aluminium	mg/kg	17 111	13 000	22 000	11 200	10 000	13 000	95
Arsenic	mg/kg	53	40	73	63	56	72	95
Cadmium	mg/kg	454	290	780	220	2	330	95
Chloride	mg/kg	145	30	145	46	30	55	95
Chromium	mg/kg	343	170	850	168	130	230	95
Copper	mg/kg	4422	3000	6700	3160	2700	3700	95
Cyanide	mg/kg	4.6	1.7	6.5	2.6	0.6	4.7	90
l.o.i.* at 105°C	%	1.6	1.3	1.9	1.6	1	3.4	
Lead	mg/kg	664	400	840	1800	1300	2500	95
Mercury	mg/kg	1.4	0.5	2.6	0.9	0.8	1.3	90
Nickel	mg/kg	4925	3100	7900	3080	2100	5300	95
pH	pH	8.2	7.7	8.5	7.6	7.5	7.7	95
Phenols	mg/kg	1.9	1	3.3	1.5	0.9	2.2	
Sulphate	mg/kg	5790	3980	7620	4644	3740	5420	90
Sulphide	mg/kg	< 20	< 20	< 20	39	< 20	64	
t.e.m	mg/kg	12 090	8170	15 300	15 400	14 000	17 200	
Total PAHs	mg/kg	94	58	129	68	51	95	
Zinc	mg/kg	1026	640	1400	1560	1400	1800	95

* l.o.i. = loss on ignition

6.3.3 Leaching tests (NRA method) on untreated materials

Samples of untreated materials A and B were subjected to the NRA leaching test (Lewin *et al*, 1994). This procedure requires crushing the larger particles and screening off those remaining that are coarser than 5 mm in size. The results of the leaching tests are given in Table 6.6. In general the untreated materials A and B, although containing substantial total amounts of the various potential contaminants (Table 6.5) were not unduly susceptible to leaching, ie those contaminants seem to be relatively unavailable. Nevertheless, the levels of contamination present an unacceptably high risk in respect of potential harm to human health (eg from ingestion). The amounts of cadmium leached from material B were above UK Drinking Water/EQS levels.

Table 6.6 *Results of NRA leaching tests on untreated materials A and B*

Determinand	Unit	Untreated material A			Untreated material B		
		Sample ref.		Means	Sample ref.		Means
		B00321	B00322		B00331	B00332	
Aluminium	mg/l	0.54	0.73	0.64	0.08	0.28	0.18
Arsenic	mg/l	<0.005	<0.005	<0.005	<0.005	<0.005	<0.005
Cadmium	mg/l	0.03	0.0044	0.002	0.0133	0.0174	0.015
Chloride	mg/l	21	5	13	8	6	7
Chromium	mg/l	<0.02	0.04	<0.03	<0.02	<0.02	<0.02
Copper	mg/l	0.06	0.06	0.06	0.03	0.03	0.030
Lead	mg/l	0.009	<0.005	<0.007	<0.005	<0.005	<0.005
Mercury	mg/l	<0.0005	<0.0005	<0.0005	<0.0005	<0.0005	<0.0005
Nickel	mg/l	<0.02	<0.02	<0.02	<0.02	<0.02	<0.02
pH	pH	7.6	7.7	7.7	7.5	7.6	7.6
Sulphate	mg/l	164	159	162	239	167	203
Sulphide	mg/l	<0.02	<0.02	<0.02	<0.02	<0.02	<0.02
t.e.m.	mg/l	12	7	10	11	8	10
Total cyanide	mg/l	<0.05	<0.05	<0.05	<0.05	<0.05	<0.05
Total phenol	mg/l	<0.05	<0.05	<0.05	<0.05	<0.05	<0.05
Zinc	mg/l	0.05	<0.02	<0.02	<0.02	<0.02	<0.02

6.3.4 Leaching tests (NRA method) on treated materials

The results from leaching tests on specimens derived from cube and core samples of the treated materials at different ages since processing are presented in Table 6.7 for treated material A and in Table 6.9 for treated material B. The ages correspond to times since processing of 28 days and 17 months for the cube samples and three, six and 15 months for the cores. In both these tables, the mean results of leach tests on untreated materials are also included. The results are presented across the tables in time sequence.

At corresponding ages to the leach testing, specimens of the cubes and cores were taken for analysis of their chemical total contents. These results are summarised in Table 6.8 for treated material A and in Table 6.10 for treated material B. Again, for comparison, the mean values of the equivalent analyses on the untreated materials A and B are also included on these tables.

The sequence of listing the determinands is the same in all four tables.

5 Immersion tests on treated materials

Intact duplicate specimens in the form of cylinders cast from the processed materials A and B were immersed in distilled water. The cylinders were 75 mm in diameter and 70 mm long. The immersion water was changed daily. The results of the tests are expressed here in Table 6.11 in terms of chemical analyses carried out on the recovered water on the first, fourth and tenth days since the cylinders were immersed.

Table 6.7 *Results of NRA leaching tests on untreated and treated material A*

Leachate determinand	Unit	Untreated mean	Treated material				
			28 day cube	3 mths core	6 mths core	15 mths core	17 mths cube
Aluminium	mg/l	0.64	0.7	20.6	85	8	1.2
Arsenic	mg/l	<0.005	<0.005	<0.005	<0.005	<0.005	<0.005
Cadmium	mg/l	0.002	<0.005	0.001	<0.005	<0.005	<0.0005
Chloride	mg/l Cl	13	11	2	6	9	12
Chromium	mg/l	<0.03	0.1	<0.02	<0.02	0.06	0.14
Copper	mg/l	0.06	0.3	0.11	0.28	0.38	0.63
Hydrocarbons by IR	mg/l	nd	nd	<0.25	<0.25	nd	nd
Lead	mg/l	<0.007	0.012	0.058	0.017	<0.005	0.007
Loss on drying at 105°C	% w/v			nd	nd	nd	nd
Mercury	mg/l	<0.0005	<0.0005	<0.0005	<0.0005	<0.0005	<0.0005
Nickel	mg/l	<0.02	<0.02	<0.02	<0.02	<0.02	<0.02
pH (20% w/v water extract)	pH	7.7	11.9	12	11.6	11.8	11.7
Sulphide	mg/l S	<0.02	<0.02	<0.02	<0.02	<0.02	<0.02
t.e.m.	mg/l	10	<5	nd	11	nd	12
Total cyanide	mg/l CN	<0.05	<0.05	<0.05	<0.05	<0.05	0.08
Total PAH	mg/l	nd	nd	<0.01	19	<5	nd
Total phenol	mg/kg	<0.05	<0.05	<0.05	0.09	0.21	0.26
Total sulphate	mg/l SO_4	162	nd	1	11	27	32
Water-soluble sulphate (1:2)	g/l SO_4	nd	nd	nd	nd	nd	nd
Zinc	mg/l	<0.02	<0.02	0.02	<0.02	<0.02	0.08

Table 6.8 *Total chemical contents of untreated and treated material A at the times of NRA leaching tests*

Determinand	Units	Untreated mean	Treated material 3 mths core	Treated material 6 mths core
Aluminium	mg/kg	17 111	13 000	26 000
Arsenic	mg/kg	53	17	8
Cadmium	mg/kg	454	47	380
Chloride	mg/kg Cl	145	10	70
Chromium	mg/kg	343	91	190
Copper	mg/kg	4422	1400	2600
Hydrocarbons by IR	mg/kg	nd	4513	1824
Lead	mg/kg	664	770	330
Loss on drying at 105°C	% w/v	1.6	5.5	4.5
Mercury	mg/kg	1.4	0.4	1.2
Nickel	mg/kg	4295	860	2400
pH (20% w/v water extract)	pH	8.2	12.1	11.6
Sulphide	mg/kg S	<20	<20	25
Toluene extractable matter	mg/kg	12 090	8940	4610
Total cyanide	mg/kg CN	4.6	0.5	0.7
Total PAH	mg/kg	94	14	nd
Total phenol	mg/kg	1.9	0.8	0.09
Total sulphate	mg/kg SO4	5790	6260	6920
Water-soluble sulphate (1:2)	g/l SO4	nd	<0.01	0.025
Zinc	mg/kg	1026	800	710

Table 6.9 *Summary results of NRA leaching tests on untreated and treated material B*

Leachate determinand	Units	Untreated mean	Treated material				
			28 day cube	3 mths core	6 mths core	15 mths core	17 mths cube
Aluminium	mg/l	0.18	1.125	18.7	51	2	1.6
Arsenic	mg/l	<0.005	<0.005	<0.005	<0.005	<0.005	<0.005
Cadmium	mg/l	0.015	<0.005	<0.0005	<0.0005	<0.0005	<0.0005
Chloride	mg/l Cl	7	9	2	4	12	6
Chromium	mg/l	<0.02	0.05	<0.02	<0.02	0.09	0.09
Copper	mg/l	0.030	0.22	0.12	0.11	0.19	0.24
Hydrocarbons by IR	mg/l		nd	<0.25	<0.25	nd	nd
Lead	mg/l	<0.005	0.047	0.042	0.93	0.02	0.015
Loss on drying at 105°C	%w/v			nd	nd	nd	nd
Mercury	mg/l	<0.0005	<0.0005	<0.0005	<0.0005	<0.0005	<0.0005
Nickel	mg/l	<0.02	<0.02	<0.02	<0.02	<0.02	<0.02
pH (20% w/v water extract)	pH	7.6	12	12	11.9	11.8	11.8
Sulphide	mg/l S	<0.02	<0.02	<0.02	<0.02	<0.02	<0.02
Toluene extractable matter	mg/l	10	6.5	nd	31	nd	21
Total cyanide	mg/l CN	<0.05	<0.05	<0.05	<0.05	<0.05	<0.05
Total PAH	µg/l		nd	<0.01	15	<5	nd
Total phenol	mg/kg	<0.05	<0.05	<0.05	0.09	0.09	0.06
Total sulphate	mg/l SO$_4$	203	nd	1	8	17	30
Water-soluble sulphate (1:2)	g/l SO$_4$		nd	nd	nd	nd	nd
Zinc	mg/l	<0.02	<0.02	0.04	0.03	<0.02	<0.02

Table 6.10 *Summary results of total chemical content on untreated and treated material B at the time of NRA leaching tests*

Determinand	Units	Untreated mean	Treated material 3 mths core	Treated material 6 mths core
Aluminium	mg/kg	11 200	16000	16 000
Arsenic	mg/kg	63	8	14
Cadmium	mg/kg	220	49	100
Chloride	mg/kg Cl	46	10	45
Chromium	mg/kg	168	83	91
Copper	mg/kg	3160	1200	1300
Hydrocarbons by IR	mg/kg	nd	3948	4608
Lead	mg/kg	1800	670	3900
Loss on drying at 105°C	% w/v	1.5	8.3	4.6
Mercury	mg/kg	0.9	0.4	0.6
Nickel	mg/kg	3080	780	1100
pH (20% w/v water extract)	pH	7.6	12.1	11.9
Sulphide	mg/kg S	39	<20	<20
Toluene extractable matter	mg/kg	15 400	7190	10 400
Total cyanide	mg/kg CN	2.6	<0.5	<0.5
Total PAH	mg/kg	68	10	nd
Total phenol	mg/kg	1.5	0.7	0.09
Total sulphate	mg/kg SO_4	4644	5900	8880
Water-soluble sulphate (1:2)	g/l SO_4	nd	<0.01	0.01
Zinc	mg/kg	1560	560	910

Table 6.11 *Results of chemical analyses on immersion water changed daily after three periods of immersion for cast cylinders*

Determinand	Units	1 day (1)	1 day (2)	4 days (1)	4 days (2)	10 days (1)	10 days (2)
Aluminium	mg/l	1.72	1.7	1.53	1.21	1.05	1.19
Arsenic	mg/l	<0.005	<0.005	<0.005	<0.005	<0.005	<0.005
Cadmium	mg/l	<0.0005	<0.0005	<0.0005	<0.0005	<0.0005	<0.0005
Chloride	mg/l	12	10	6	5	2	2
Chromium	mg/l	0.08	0.08	0.03	0.03	<0.02	<0.02
Conductivity 20°C	µS/cm	526	547	332	529	384	304
Copper	mg/l	0.13	0.15	0.06	0.06	<0.02	0.03
Lead	mg/l	<0.005	<0.005	<0.005	<0.005	<0.005	<0.005
Mercury	mg/l	<0.0005	<0.0005	<0.0005	<0.0005	<0.0005	<0.0005
Nickel	mg/l	<0.02	<0.02	<0.02	<0.02	<0.02	<0.02
pH	pH	11.1	11.1	11	11.3	11.2	11
Sulphate	mg/l	23	19	5	3	1	1
Total cyanide	mg/l	<0.05	<0.05	<0.05	<0.05	<0.05	<0.05
Zinc	mg/l	<0.02	<0.02	<0.02	<0.02	<0.02	<0.02

(1) Immersion water treated material A   (2) Immersion water treated material B

6.3.6 Surface waters and the compacted slab at Wombwell

At the Wombwell site, the treated materials were placed and compacted by roller to form a single monolithic slab, but with each of the treated materials A and B placed in separate parts of the slab. Since then (ie for over five years) the slab has been exposed to the weather, with the potential for infiltration and erosion by rainwater. In order to establish the susceptibility of the compacted treated material of the slab to release contamination, three aspects were monitored.

In the first twelve months, samples were taken of:

- runoff water from the slab surface
- water from the collector drain below the slab, ie water percolating through the slab
- sediments downstream of the slab in the brook that passes through the site and which receives the runoff water.

The three sets of analyses each for two times are presented in Table 6.12. The summer of 1995 was hot and dry, ie evaporation would have been higher than usual, which could have resulted in relatively high concentrations of leached and eroded matter accumulating in the collection sumps.

When the TRL study was undertaken in 1998–1999, the slabs were inspected and further water samples obtained and analysed (see Table 6.13).

Table 6.12 *Analyses of runoff and percolation waters and downstream sediments at the site of the slab compacted from treated materials A and B*

Determinand	Run-off 3 mths (mg/l)	12 mths (mg/l)	Percolation 3 mths (mg/l)	12 mths (mg/l)	Receiving sediments Pre-trial (mg/kg)	12 mths (mg/kg)
Aluminium	0.12	0.16	0.65	0.29		
Arsenic	<0.005		<0.005		25	42.6
Cadmium	<0.0005		<0.0005		3.13	1.66
Chloride	6		6			
Chromium	0.09		0.08		111	108
Copper	0.32		0.2		126	12
Cyanide	<0.05		<0.05			
Lead	0.007	<0.1	<0.005	<0.1	155	139
Mercury	<0.0005		<0.0005		0.39	0.43
Nickel	0.07		<0.02		103	85
Phenol	0.07	<0.15	<0.05	<0.15		
Sulphide	<0.02		<0.02			
t.e.m.	13		10			
Zinc	0.04	<0.1	<0.02	<0.1	431	432

Table 6.13 *Inorganic contaminants in drain water samples collected in TRL study (from Board and Reid, 1999)*

Determinand	Drain 1 (mg/l)	Drain 2 (mg/l)
Al	<0.02	0.03
As	<0.005	<0.005
B	0.57	0.39
Ca	32.5	34.7
Cd	<0.0005	<0.0005
Conductivity μs/cm at 20°C	300	450
Cr	0.017	0.025
Cu	0.04	<0.02
Fe	<0.02	<0.02
Hg	<0.0005	<0.0005
K	15.3	40.8
Mg	3.92	2.51
Mn	<0.02	<0.02
Na	9.75	22.9
Ni	<0.02	<0.02
Pb	<0.005	<0.005
pH	8.1	8.3
Se	0.03	0.02
Total CN	<0.05	<0.05
Zn	<0.02	<0.02

7 Subsidiary trials

7.1 SCOPE OF THE WORK

Four materials were brought to the Wath processing site for treatment by the *Geodur* system: three industrial by-products and a contaminated soil.

1. Material C, a lagoon-dried pulverised fuel ash from coal combustion at Eggborough power station in North Yorkshire.

2. Material D, a zinc-smelting slag.

3. Material E, an intermediate (vertical retort) residual from zinc smelting.

4. Material F, a contaminated soil from the Derby Pride City Challenge development site.

Each of these materials was sampled and tested and an appropriate mix design formulated for it with the *Geodur* Tracelok™ additive (see Section 7.3). Processing took place immediately after the main demonstration. The processed materials were transported to the slab site at Wombwell and cast in place as separate slabs (each about 8–10 m by 4 m in plan area) and were compacted by the same vibrating roller as used for the slab made of the main trial materials.

Cube samples were made from the processed materials for determination of compressive strengths, chemical contents and leachability. Cores were taken from the trial slabs at the same time as for the main slab, ie at three, six, 12 and 15 months; these were also tested for compressive strength and chemical content and leachability.

Table 7.1 *Particle size distributions of subsidiary trial materials C, E and F*

Sieve size	Percentage passing by dry weight for material		
	C	E	F
50 mm			100
20 mm		100	98
2 mm	100		80
1.2 mm	98	61	
600 µm	96	21	30
300 µm	92.5		
150 µm	73		
75 µm	40		2

7.2 ENGINEERING AND CHEMICAL DESCRIPTIONS

7.2.1 Material C

Material C, a pfa, is a light grey, silty-fine ash. Its grading, with those of the subsidiary trial materials, is given in Table 7.1. The natural moisture content of the material was 22 per cent. Information provided by the supplier about the chemical composition of this material is presented in Table 7.2.

Table 7.2   *Chemical composition of material C*

Determinand	Proportion by mass (%)	Determinand	Value
SiO_2	52.1	Water-soluble chloride	< 0.1 g/l
TiO_2	0.87	Water-soluble sulphate	< 0.1 g/l
Al_2O_3	23.9	Loss on ignition	5.93 % w/w
Fe_2O_3	7.7		
MgO	1.66		
BaO	0.11		
CaO	1.56		
Na_2O	2.24		
K_2O	3.52		

7.2.2 Material D

Material D, a zinc-smelting slag, is a black glassy grit, of predominantly sand-size particles. The grading is given in Table 7.1. The natural moisture content was 2.7 per cent. The chemical composition of this material, given in Table 7.3, is as provided by the supplier.

Table 7.3   *Chemical composition of material D*

Determinand	Proportion by mass (%)	Determinand	Proportion by mass (%)
Al_2O_3	7.5	MnO	1.2
As	0.9	Pb	1.3
BaO	0.3	S	2
CaO	14	Sb	0.09
Cu	0.67	SiO_2	17.8
FeO	40	Sn	0.03
MgO	1.6	Zn	8.9

7.2.3 Material E

Material E, the intermediate residual from a vertical retort in zinc smelting, is a grey ash, but containing weakly cemented aggregations like briquettes. The grading is given in Table 7.1. The natural moisture content of the material was about 50 per cent. The only information provided about the chemical composition of this material was that lead concentrations are of the order of 1.5–2 per cent.

7.2.3 Material F

Material F is a made ground from a former gasworks site. It is largely a dark brown silty clay, but with an oily appearance and containing debris (plastic bags, bricks, electric cables, etc). Not only did the soil have an oily appearance, it had an acrid smell. The material's grading was not determined, the contents of the bulk sample being considered too heterogeneous, nor was a particle size distribution provided by the source site. Screening of this material left a poorly sorted mixture of clay lumps of 25–150 mm in size including fragments of other materials. Moisture contents were not measured, nor was the chemical composition of this material provided by the supplier.

7.3 MIX DESIGN CONSIDERATIONS

For the subsidiary trial materials, a formal mix design process was not practicable. The aim was to use the material with the addition of 10 per cent of cement, a similar amount of aggregate to that with the main trial materials A and B, and an appropriate amount of water for hydration and compaction. A smaller amount of aggregate was used with material D. The Tracelok™ additive was batched at 1.5 litres per tonne of dry soil. Table 7.4 shows the treatment mixes of the subsidiary trial materials.

Table 7.4 Treatment mixes of subsidiary trial materials

Material	C	D	E	F
Type	PFA	Zinc-smelting slag	Intermediate residual from zinc smelting	Made ground from gasworks site
As received moisture content*	20.2	3	14.4	20.2
Cement*	10	10	10	10
Geodur additive*	0.15	0.15	0.15	0.15
Aggregate content*	18	10	18	18
Untreated waste material as a proportion of dry mass of whole mix	72	80	72	72

* As percentage of dry mass of untreated material.

7.4 COMPRESSIVE STRENGTHS

The compressive strengths determined at different times are summarised in Table 7.5 for the on-site mixed cubes, and in Table 7.6 for the specimens prepared from 143 mm diameter core samples taken from the cast-in-place slabs at ages up to 12 months. Also included are the data from Board and Reid (1999) of their tests at 40 months on three no. 97 mm diameter cores taken from the slabs of each treated material. Note that the results for the core samples have not been converted to equivalent cube strengths.

Table 7.5 *Compressive strengths of cube samples after two curing periods*

Treated material	Compressive strength (N/mm²) at	
	28 days	12 months
C	2.0 (air cured)	10 (saturated)
D	not determined	not determined
E	4.5 (air cured)	5 (saturated)
F	1.5 (air cured)	4 (saturated)

Table 7.6 *Compressive strength of cores taken at different times after casting*

Material	Compressive strength (N/mm²) at			
	3 months	6 months	15 months	40 months
C	10.5	13.3	15.5	9.8–16.8
D	13.0	16.7	19.0	10.2–17.3
E	4.5	7.0	7.5	4.0–14.68
F	2.5	3.6	1.5	2.7–5.8

7.5 LEACHING TESTS (NRA METHOD) ON SUBSIDIARY TRIAL MATERIALS

Summaries of the analytical test results and other data relevant to the results of the leaching tests on treated materials C, D, E and F are given in Tables 7.7–7.10.

Table 7.7 *Treated material C: results of leaching tests (NRA method)*

Determinand	Units	Leachate composition from sample of age and type				
		1 mth cube	3 mths core	6 mths core	15 mths core	17 mths cube
Chloride	mg/l	14	2	3	9	5
Hydrocarbons by IR	mg/l		<0.25	<0.25		
pH	pH	12.2	12	12.1	11.7	11.8
Sulphate	mg/l		1	7	22	12
Sulphide	mg/l	<0.02	<0.02	<0.02	<0.02	<0.02
t.e.m.	mg/l	5	–	47	7	10
Total aluminium	mg/l	0.89	15.3	4.2	4.5	2.0
Total arsenic	mg/l	<0.005	<0.005	<0.005	<0.005	<0.005
Total cadmium	mg/l	<0.0005	<0.0005	<0.0005	<0.0005	<0.0005
Total chromium	mg/l	0.04	<0.02	0.03	0.06	0.06
Total copper	mg/l	<0.02	<0.02	<0.02	0.02	0.03
Total cyanide	mg/l	<0.05	<0.05	<0.05	<0.05	<0.05
Total lead	mg/l	0.02	<0.005	0.03	<0.005	0.005
Total mercury	mg/l	<0.0005	<0.0005	<0.0005	<0.0005	<0.0005
Total nickel	mg/l	<0.02	<0.02	<0.02	<0.02	<0.02
Total PAHs	mg/l		<0.01	0.002	nd	nd
Total phenols	mg/l	<0.05	<0.05	<0.05	0.21	<0.05
Total zinc	mg/l	<0.02	<0.02	<0.05	<0.02	<0.02

Table 7.8 *Treated material D: results of leaching tests (NRA method)*

Determinand	Units	Leachate composition from sample of age and type				
		1 mth cube	3 mths core	6 mths core	15 mths core	17 mths cube
Chloride	mg/l	9	3	7	14	6
Hydrocarbons by IR	mg/l		<0.25	<0.25		
pH	pH	11.6	12.1	12	11.8	11.2
Sulphate	mg/l		<1	40	34	244
Sulphide	mg/l	<0.02	<0.02	<0.02	<0.02	<0.02
t.e.m.	mg/l	9	–	31	<5	19
Total aluminium	mg/l	0.98	5.7	0.8	0.53	0.79
Total arsenic	mg/l	<0.005	<0.005	<0.005	0.005	<0.005
Total cadmium	mg/l	<0.0005	<0.0005	<0.0005	<0.0005	<0.0005
Total chromium	mg/l	<0.02	<0.02	<0.02	<0.02	<0.02
Total copper	mg/l	0.18	0.23	0.28	0.23	<0.02
Total cyanide	mg/l	2.5	0.11	0.87	0.81	7.88
Total lead	mg/l	0.017	0.012	0.04	0.002	0.005
Total mercury	mg/l	<0.0005	<0.0005	<0.0005	<0.0005	<0.0005
Total nickel	mg/l	0.03	0.07	0.09	0.08	<0.02
Total PAHs	mg/l		<0.01	0.014	nd	nd
Total phenols	mg/l	<0.05	0.11	0.25	0.52	0.35
Total zinc	mg/l	0.04	0.04	0.03	<0.02	<0.02

Table 7.9 *Treated material E: results of leaching tests (NRA method)*

Determinand	Units	Leachate composition from sample of age and type				
		1 mth cube	3 mths core	6 mths core	15 mths core	17 mths cube
Chloride	mg/l	8	2	6	9	5
Hydrocarbons by IR	mg/l		<0.25	<0.25		
pH	pH	12	11.8	11.8	11.4	11.4
Sulphate	mg/l		3	50	71	105
Sulphide	mg/l	<0.02	<0.02	<0.02	<0.02	<0.02
t.e.m.	mg/l	9	–	10	5	16
Total aluminium	mg/l	3.06	19.6	4.9	4.1	2.75
Total arsenic	mg/l	<0.005	<0.005	0.017	0.016	0.017
Total cadmium	mg/l	<0.0005	<0.0005	<0.0005	<0.0005	<0.0005
Total chromium	mg/l	0.03	<0.02	0.06	0.07	0.10
Total copper	mg/l	<0.02	0.02	<0.02	0.03	0.03
Total cyanide	mg/l	<0.05	<0.05	<0.05	<0.05	<0.05
Total lead	mg/l	0.095	0.061	0.06	0.018	0.015
Total mercury	mg/l	<0.0005	<0.0005	<0.0005	<0.0005	<0.0005
Total nickel	mg/l	<0.02	<0.02	<0.02	<0.02	<0.02
Total PAHs	mg/l		<0.01	0.001	nd	nd
Total phenols	mg/l	<0.05	<0.05	<0.05	0.06	<0.05
Total zinc	mg/l	0.02	0.09	0.03	<0.02	<0.02

Table 7.10 *Treated material F: results of leaching tests (NRA method)*

Determinand	Units	Leachate composition from sample of age and type				
		1 mth cube	3 mths core	6 mths core	15 mths core	17 mths cube
Chloride	mg/l	No available data		3	111	No available data
Hydrocarbons by IR	mg/l		<0.25	<0.25		
pH	pH		11.8	12.2	12.2	
Sulphate	mg/l		3	9	6	
Sulphide	mg/l		<0.02	<0.02	<0.02	
t.e.m.	mg/l		–	14	<5	
Total aluminium	mg/l		19.6	0.77	0.37	
Total arsenic	mg/l		<0.005	0.06	<0.005	
Total cadmium	mg/l		<0.0005	<0.0005	<0.0005	
Total chromium	mg/l		<0.02	<0.02	<0.02	
Total copper	mg/l		0.02	0.11	0.03	
Total cyanide	mg/l		<0.05	<0.05	<0.05	
Total lead	mg/l		0.061	4.53	0.74	
Total mercury	mg/l		<0.0005	<0.0005	<0.0005	
Total nickel	mg/l		<0.02	0.03	<0.02	
Total PAHs	mg/l		<0.01	0.005	nd	
Total phenols	mg/l		<0.05	<0.05	0.08	
Total zinc	mg/l		0.09	0.07	<0.02	

8 Operational good practice

8.1 PRINCIPLES OF GOOD PRACTICE

Solidification/stabilisation is one of many technologies that can be applied to the remedial treatment of contaminated soils. As with all approaches to remediation, good practice in the use of this technique on a contaminated site should achieve:

- regulatory compliance, particularly in relation to matters affecting human health and safety and the environment
- consistency between the various methods applied to carrying out the work
- adherence to the principles and procedures of quality assurance and quality control in carrying out and in recording the work
- transparency of overall approach and in how to test and monitor the work.

These are broad objectives, which have to be applied to the many component activities specific to the particular remedial technique and situation in which the remediation is taking place. Generic guidance about the types of tasks involved in selecting a remedial treatment method and about the application of solidification/stabilisation to remediation is given in CIRIA Special Publications 104 and 107 (Harris, Smith and Herbert, 1995 a and 1995 b) and in Evans *et al* (in press).

8.2 PLANNING AND ORGANISATION

In addition to the scheduling of the treatment process in relation to the contaminated material, other tasks will have a bearing on the overall programme planning. While these will vary with the specific application, they are likely to be on the critical path, namely:

- regulatory approvals
- feasibility and treatability testing
- the mobilisation of equipment and resources
- sampling and monitoring programmes (before, during and after treatment)
- third party validation.

Even though the demonstration involved small quantities of material, small plant and few personnel, the organisational structure which developed can be applied to much larger works. Thus the experience gained in carrying out the trials, together with that of *Geodur* from practice in Europe, can be represented by the organisation chart shown in Figure 8.1 (which also represents information flow). That figure is essentially the organogram given earlier as Figure 4.2, but made non-specific and extended to include client, regulators and, in relation to the CDM regulations, the planning supervisor.

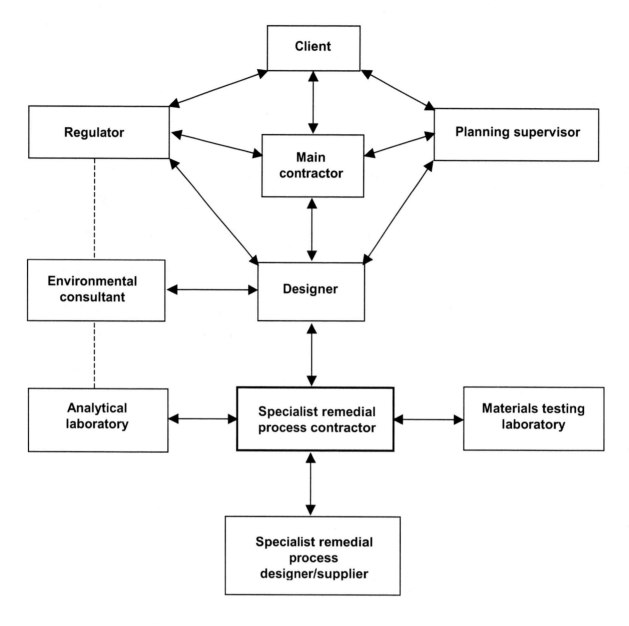

Figure 8.1 *Organisational structure*

8.3 OPERATIONS

As discussed below, the demonstration revealed various operational matters which would have to be addressed in any commercial application of this technology in processing contaminated ground.

8.3.1 Pre-processing

The screening of the excavated material to remove debris, such as bricks and concrete fragments, was successful in producing relatively homogeneous material in stockpiles ready for loading into the *Geodur* process. Reducing the average size of the material still further as an additional stage of the pre-processing, ie breaking up lumps and larger particles, would make for more efficient mixing. This would improve the quality of the processed material both physically and in terms of its environmental performance.

The necessary degree of pre-processing depends not only on the nature of the soil to be treated (ie its friability, plasticity, moisture content, etc) and the coarse material within it in relation to the mixing process, but also on its end-use and the required performance criteria, eg the stringency of the specified compliance testing.

8.3.2 Processing

The batching plant used in the trials was a smaller model than had originally been intended for the demonstration, that equipment being unavailable at the time of the demonstration. The batching plant of the trial would not be able to process the much larger quantities of material as efficiently and accurately as would be required for a remediation contract. It was also more restricted in the adjustments that could be made to batching, and this somewhat constrained mix design. The general opinion, therefore, was that because this equipment did not produce as high a quality of processed material as would the other plant, it reduced the effectiveness of the demonstration.

The batching system recommended for use with a treatment such as the *Geodur* system should have the following attributes:

- weigh belts with computer control over feed rates, and volumes
- high-energy mixing with control over resistance time
- a minimum of two vibrating hoppers for materials
- covered conveyors or secondary collection of any spilled material.

For a medium to large on-site remediation project, one or more batching plants with a throughput of not less than 150 tonnes per hour would be required. Mixing the various components that make up the processed material is critical to meeting physical and environmental performance criteria, independent of the exact nature of those criteria. A high-energy continuous or batch mixer is therefore necessary.

In addition, full on-line control of the volume of each component is required. This can usually be achieved with multiple hopper feeds and computer-controlled electronic weigh belts. A continuous record of all volumes and feed rates is required for construction or process quality assurance and control (CQA/PQA).

8.4 SAMPLING AND TESTING

The sampling and testing regime designed for this demonstration raised a number of points of good practice that would need to be addressed in any full commercial application.

8.4.1 Characterisation of the in-ground materials or other materials for processing

The as-excavated or as-supplied materials to be processed would have to be subjected to the same leaching test as the processed *Geodur* material. Any categorisation of these materials should be made on the basis of leachable fractions and not of total chemical contents.

All other mix components, typically cementitious binder and aggregates, to be used in the process to form the *Geodur* material should be subject to the same leaching tests as the contaminated materials.

Samples of the contaminated materials and the other added components should be taken at regular intervals throughout the processing period and at times when any changes are made to either the process materials or to the process conditions.

Total chemical content determinations are also required as baseline values and for comparisons with the leachable fraction.

For both physical and chemical tests, a sufficient number of tests should be carried out to enable the range and mean values of each parameter to be established.

9 Discussion of results

The following comments are largely based on the assessment made by Board and Reid (1999) of the initial testing results and their own later more extensive study of the treated materials. Thus the discussion refers to some of their results on the aged materials.

The following aspects of the performance and properties of the solidified/stabilised trial materials are considered, each of which is discussed below:

- contaminants in the untreated and treated materials
- leaching test results: leached inorganics
- leaching test results: leached organics
- permeability and leachability of the compacted treated material
- strength
- durability.

9.1 CONTAMINANTS IN THE UNTREATED AND TREATED MATERIALS

All of the wastes can be considered as containing contaminants. Materials A and B were initially classified as special and hazardous wastes. The other wastes brought for treatment contained a range of inorganic and organic contaminants such that, in the case of metals, the mean content values exceeded the ICRCL thresholds for use in open spaces and for buildings. All the treated materials (other than C) exceeded these ICRCL guideline values for at least three metals; and material C did so for copper and nickel.

In the treatment process, these contaminants remain – except if there are losses as dust, in lost process water and by volatilisation. The chemicals become less mobile in the solidified material. While the total amount of each original contaminant should therefore remain essentially the same, its concentration as a proportion of the total mass of material is reduced to the extent that additional material, such as cement and aggregate, has been added. If as here the added materials contain elements (eg aluminium in cement) and compounds that are the same as those analysed for the waste itself, the concentrations reported in the analytical results will increase. This appears to be the case for aluminium. The composition of the Tracelok™ additive is not known.

The addition of cement raised the pH of the treated materials up to 11.5–12, and this was reflected in the pH values of the leachates from the leaching tests and on the site.

9.2 LEACHING TEST RESULTS: LEACHED INORGANICS

The metals Cd, Cu, Hg, Ni, and Zn, as measured by the concentrations in the leachates, were effectively immobilised below the environmental quality standards derived from drinking water standards. In most cases these concentrations were lower for the treated materials than for the untreated wastes.

Aluminium, however, was leached above its environmental quality (drinking water) standard for all the untreated and treated materials. Lead contents in the leachates for treated materials A, B, D and E and chromium contents for A, B, C and E at some stages in the study were also above the standards value. Note that the chromium content in leachates of the untreated materials were lower than the environmental quality standard.

9.3 LEACHING TEST RESULTS: LEACHED ORGANICS

The three wastes containing organics (petroleum and diesel range hydrocarbons, PAHs, PCBs and phenols) were A, B and F. In none of the leachates from tests on these untreated materials were the organics contents greater than the drinking water standards (although they were not analysed for total PAH contents). In the leachate tests undertaken in the CIRIA part of the trial on the treated materials, all the hydrocarbon contents were below the detection limit of 0.25 mg/l of the analytical method used. Note that the environmental quality standard for drinking water is less than this, ie 0.01 mg/l. Subsequent tests by TRL with a much lower detection limit (10 µg/l) were all below the quality standard for hydrocarbons.

Higher concentrations of phenols were measured on the tests leachates of the treated materials than of the untreated materials. Also the TRL tests measured total PAH concentrations from test leachates on the aged treated materials to be higher than the drinking water standards.

9.4 PERMEABILITY OF THE COMPACTED TREATED MATERIAL

The TRL study measured the porosities and coefficients of hydraulic conductivity of cores of the treated materials. The porosities ranged from 23–26 per cent for treated materials A, B, D and F, and were 32 and 34 per cent for treated materials C and F, these values reflecting the overall relation grading and compactability.

The hydraulic conductivities of treated materials A and B were about 1.5×10^{-9} m/s. Treated material C had the highest value at 2.3×10^{-7} m/s and treated material E the lowest at 8.1×10^{-10} m/s, the other two being of the order of 10^{-8} m/s. These are all relatively impermeable, particularly those in the 10^{-9} range, so any penetration and percolation of water would be slow.

9.5 STRENGTH

Although there was considerable difference in the ranges of compressive strengths achieved by the treated materials and – for those with organic contents and high metal concentrations the early gain in strength with time was slow– they all eventually achieved substantial strengths in place. Thus on this criterion, there could be potential with some of the treated wastes for use as a lower pavement layer. PFA (material C) can be pozzolanic and is used as a cement replacement in concrete mixes. Its strength when treated with cement was expected to be high.

9.6 DURABILITY

While no specific tests were carried out as indicators of durability, the slabs were left exposed to the weather over four winters without protection. The surfaces were fractured and disturbed, it is assumed by frost action.

10 Concluding remarks

1. In the field trial and demonstration of the *Geodur* system for solidification/ stabilisation, cement and aggregate were mixed with two contaminated slags from the Wath recycling site and four wastes from elsewhere together with mix water and an additive. The treated materials were readily laid and compacted into ground slabs. The slabs were left exposed to the weather, ie without surface protection for over four years since they were laid in December 1994. The testing and analytical studies carried out by CIRIA and later by the Transport Research Laboratory in 1998–1999 are consistent.

2. The full-scale demonstration of the *Geodur* system was successful in that contaminated materials were rendered to a condition in which they posed less threat to people and the environment. Some very poor quality wastes were transformed into viable, relatively strong and durable construction materials, that might be used, for example, as a sub-base or lower pavement layer to a hardstanding or similar.

3. Routine plant, but of relatively small capacity was used but with larger treatment volumes, economies of scale and greater control could be obtained with higher capacity equipment. For the particular wastes of the trial, their early strengths even with 10 per cent cement were relatively low. If higher cement contents are needed, the cost might preclude their use.

4. All the samples can be considered as contaminated. The *Geodur* treatment was effective in solidifying all of them, ie creating a bound material. The treatment as a process of gain in compressive strength progressed at different but steady rates in the different materials. Even the weakest treated material F had strengths of at least 2.6 N/mm^2 after 1216 days.

5. Whether there is effective immobilisation of the contaminants is less clear. The immobilisation of contaminants is measured by comparing leachate results of treated and untreated materials. Where reduced concentrations of particular contaminants were measured in leachate from treated as opposed to untreated material this could be partly attributed to the diluting effect of adding aggregate and cement (28 per cent of the dry mass of treated materials A and B). On the other hand, greater concentrations of aluminium were measured in the test leachate from the treated materials, which is attributed to its being leached from the cement.

6. Nevertheless the analyses of leachate from the cast slab and from runoff water at the slab site showed that they were generally within UK drinking water standards other than for aluminium in the first study phase and for selenium and potassium in the later TRL study which were slightly over the limits. A general lesson for practice that stems from this is that it is not sufficient only to test leachates for typical contaminants of concern, such as, for example, those of concern for human health by ingestion or inhalation, but that the analytical suite should cover all contaminants which might be of concern to surface waters and groundwater, eg aluminium.

7. It is not clear what the effect the *Geodur* additive had as there were no control tests against which to compare (ie strength of materials treated with the same amounts of aggregate and cement but without the Tracelok™ additive). It is assumed, however, that it assisted stabilisation in the presence of potentially inhibiting organic compounds. Without knowing the chemical composition of the additive it is not possible to comment on its effect, if any, on fixation of the metals, nor to separate its contribution to resisting leaching or becoming part of the leachate itself.

References

BOARD, M J and REID, J M (1999)
The effects of age on cement stabilised/solidified contaminated materials
Unpublished Project Report PR/CE/135/99 E199D/HG
Transport Research Laboratory, Crowthorne, Berks

EVANS, D, JEFFERIS, S A, THOMAS, A O and CUI, S (in press)
Remedial processes for contaminated land: principles and practice
Publication C549
CIRIA, London

HARRIS, M R (1996)
Framework protocol for reporting the demonstration of land remediation technologies
Project Report 34
CIRIA, London

HARRIS, M R, SMITH, M A and HERBERT, S M (1995 a)
Remedial treatment for contaminated land, Volume IV: classification and selection of remedial methods
Special Publication 104
CIRIA, London

HARRIS, M R, SMITH, M A and HERBERT, S M (1995 b)
Remedial treatment for contaminated land, Volume VII: ex-situ remedial methods for soils, sludges and sediments
Special Publication 107
CIRIA, London

LEWIN, K, BRADSHAW, K, BLAKEY, N C, TURELL, J, HENNINGS, S M and FLAVIN, R J (1994)
Leaching tests for assessment of contaminated land: interim NRA guidance
R and D Note 301
National Rivers Authority, Bristol